Latrece Johnson, MEd, MS

WHAT IS A LIONFISH?

"Local Science by A Local Scientist" series

Dedicated to all who live on and love our waters.

--LJ

What is a lionfish?

The Lionfish, a meat eating fish native to the Pacific Ocean, is now an invasive species in the Atlantic Ocean.

Lionfish live in the waters near Pensacola, Florida.

Other names for **lionfish** include zebrafish, turkeyfish, butterfly cod, peacock lionfish, red firefish, and scorpion volitans.

Lionfish have eighteen poisonous spines along their backs and sides.

Lionfish eat native fish like grouper and snapper.

Lionfish have no known natural predators in northern Gulf waters.

Lionfish typically avoid baited hooks.

Lionfish swim deeper than scuba divers.

Lionfish eat and breed in waters too deep for expert divers to reach, also.

Female lionfish lay millions of eggs a year.

Annual tournaments and awareness festivals are held in the state of Florida to capture lionfish.

More than 15,000 lionfish were removed from Florida waters in 2018 due to these events.

Words to Know
Lionfish
Invasive
Native
Florida

Questions for Understanding

1. What is a lionfish? (p6)
2. What are some other names for lionfish you learned in this book? (p10)
3. How many spines do lionfish have on their backs? (p12)
4. What do lionfish eat? (p14)

Lionfish Craft

MATERIALS
Clean, empty water bottle
Yellow acrylic paint
Brown acrylic paint
Paintbrushes, 2
Wiggle eyes, 2
Scissors
Glue
*an adult should be present for this craft

Use the scissors to cut fins on the side of the water bottle. Use scissors to cut the base of the water bottle into spines. Use paintbrushes to add yellow paint to the third top of the bottle and one half of the bottom of the bottle; and brown paint to the middle section and remaining bottom half of the bottle. Glue the wiggle eyes onto the top of the bottle. Enjoy your creative lion fish!

Photo credit: Pinterest

Video Information about Lionfish

How Florida is handling invasive lionfish - YouTube.
https://www.youtube.com/watch?v=CSd7pgvOV3M

Venomous Lionfish | Oceans | BBC.
https://www.youtube.com/watch?v=XqGhsMhZtF0

Divers Fight the Invasive Lionfish | National Geographic.
https://www.youtube.com/watch?v=GzaeYzAC8Ro

Lionfish Articles

Coldeway, D. (2014). Sixth-grader's lionfish science project stuns conservation experts. Retrieved from https://www.nbcnews.com/science/environment/sixth-graders-lionfish-science-project-stuns-conservation-experts-n161391

Kearney, M. (2018). 15,000 lionfish removed from Florida waters, tournaments target invasive species. Retrieved from https://www.naplesnews.com/story/news/environment/2018/05/23/tournaments-help-remove-15-000-lionfish-florida-waters/635024002/

National Ocean Service. What is a lionfish? Retrieved from https://oceanservice.noaa.gov/facts/lionfish-facts.html

Please note that internet links worked at time of book publication.

Photo Credits:

Bing via Creative Commons license

www.ingramcontent.com/pod-product-compliance
Lightning Source LLC
Chambersburg PA
CBHW051824210526
45473CB00005B/1730